Aquaponics Made Easy:

A Simple and Easy Guide to Raising Fish and Growing Food Organically in Your Home or Backyard

By: Brian Grant

Published by:

Brian Grant and Random Technologies

4409 HOFFNER AVENUE, 347

Belle Isle, FL 32812

Get FREE Books at:
www.SparrowPublications.com

Disclaimer

Table of Contents

Introduction to Aquaponics

Aquaponics is a system of gardening that combines the practices of 'aquaculture' and 'hydroponics'.

Aquaculture is the raising of aquatic animal life such as: prawns, crayfish, snails, and fish, in a fish tank.

Hydroponics is the system of cultivating plants in water.

Both these systems of producing fish and produce are efficient, however, they both have a downside.

Raising aquatic animals requires that the water be constantly refreshed to remove effluent, that if left will kill the fish. Growing fruit and vegetables without soil, means that the plants require a separate supply of nutrients to grow and flourish.

By combining the two systems, the effluent from uneaten feed or from the fish themselves can be used to provide essential nutrients for plant growth. The 'clean' water is then drained back into the fish tank, thereby cutting out the need for constant fresh water. The waste product of one system becomes the very thing that the other system needs most.

P.S. Sign up for updates of our new books, free bonuses and more... Please subscribe for here:

www.SparrowPublications.com

History of Aquaponics

As with all great systems, aquaponics has been around for a long time. We can trace the use of aquaponics back to the Aztecs, who used networks of canals and stationery artificial islands on which to grow crops, using nutrient rich mud and water from the canals.

The ancient Chinese also used a system whereby finfish, ducks, catfish and plants co-existed together.

However, it is likely that the system of aquaponics that we know of today came from the New Alchemists.

In 1980, the first closed system of aquaponics was created, using effluent from fish to irrigate tomato and cucumber plants, grown in sandbags. Water from the sandbags then trickled back down into the fish tanks.

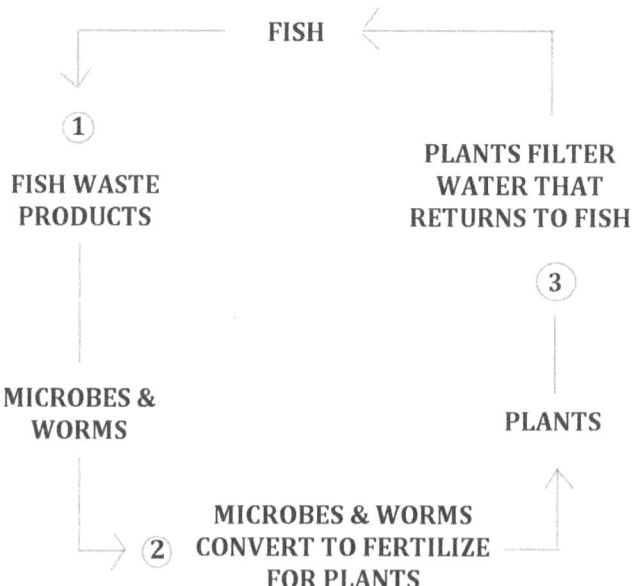

FISH

① FISH WASTE PRODUCTS

PLANTS FILTER WATER THAT RETURNS TO FISH

③

MICROBES & WORMS

PLANTS

② MICROBES & WORMS CONVERT TO FERTILIZE FOR PLANTS

Benefits of Aquaponics

There are many benefits to using aquaponics:

- Less water used than traditional gardening.

- No need for fertilizer or pesticides.

- No soil required so less labour intensive, i.e. no weeding or tilling.

- Drought proof as no need to water during summer months.

- Plants grow faster due to the constant nourishment and water.

- High crop yield due to the constant nourishment and water, up to 4 times as much.

- No waste as all waste products are used within the system.

- Sustainability as little water is required after the initial start-up.

- Low maintenance, due to the recirculation of water, the fish tank requires little cleaning.

- Saves space, due to the high crop yield, not as many crops need to be grown to acquire plentiful produce.

- Saves money. Aquaponics used less water, no fertilizers or pesticides and gives a higher return on crop yield, all saving you money.

How to Get Started

Flood and Drain System

The most common and easiest aquaponic system to use is the 'flood and drain system'. It is easy to understand and build, and so makes it the perfect place for most beginners to aquaponics to start.

It is a basic system and works on a 1:2 fish tank volume/grow bed volume

FLOOD AND DRAIN SYSTEM

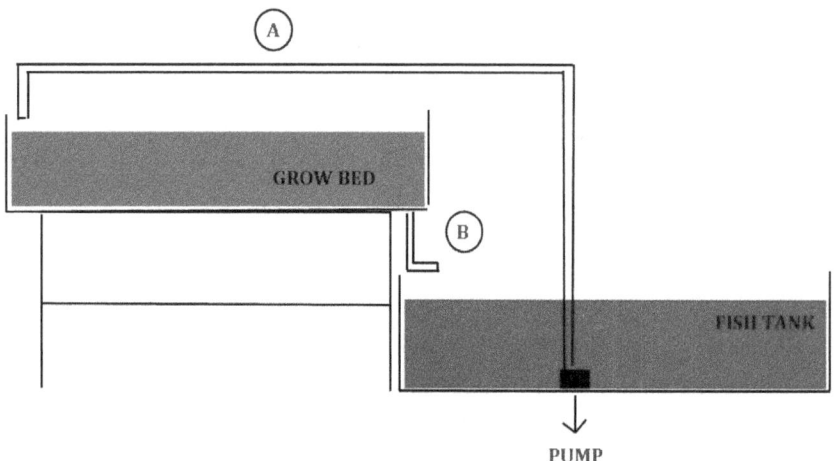

4

The key points of a flood and drain system are:

- You require a media based grow bed to be placed above the fish tank.

- Water is pumped from the fish tank to the grow bed. (A)

- The effluent rich waste drains, via gravity, down through the plants and back into the fish tank. (B)

Components of a Flood and Drain System

The flood and drain system is very easy to set up and all components can be purchased from a reliable aquaponics/aquatic store, a good DIY center or online.

You will need:

- Fish tank.
- Growing container.
- Table or platform to place the growing container on so it is higher than the fish tank.
- Piping for water flow.
- Pump/timer.
- Grow media.

The system is a simple circuit (as explained below), of water out of the fish tank to the grow bed, filtered through the grow bed and then returning to the fish tank.

How to Achieve the Flood and Drain Effect

There are two ways in which to run a flood and drain system.

1. Timer based.
2. Bell or auto siphons.

The timer based option is the easiest to run and set up. A timer is attached to the pump and set to run for 15 minutes in every 60 minutes.

During these 15 minutes, water is pumped into the bottom of the grow bed via the water pipe, until the water level reaches the level of the overflow drain.

TIP – the overflow drain should be set 1 inch below the surface of the grow bed, so as to avoid the lower leaves of the plants from becoming moldy. This then prevents algae from forming on the wet media surface.

After 15 minutes of flooding, the pump switches off and the water drains back through the water pipe at the base of the grow bed and into the fish tank.

In the second option, the 'bell or auto siphon' method, the pump is constantly on, thus sending water to the grow bed.

As the water level rises, it fills the siphon located in the grow bed. When the water reaches a set height, it overflows into a pipe within the siphon which then triggers the water in the grow bed to begin to drain back into the fish tank.

When the siphon is almost empty, air will enter which stops the draining process. The grow bed will then begin to fill up with water again.

Location and Size

One of the first things you need to do when setting up your aquaponics system is to decide where you are going to site it.

There are a few things that you need to bear in mind before doing this:

- Do you live in a climate that is temperate enough to grow your produce outside?
- If not, do you have access to a greenhouse or indoor area where your plants will receive adequate sunlight to grow, or will you need to use grow lamps?
- Remember that once sited, your system will be very difficult to move. Make sure you take time to think about the best place for your fish and produce before you start.

Once you know where you are going to site your system, you need to start thinking about the size.

Whilst aquaponics is an easy system to use, it is advised that you don't get too carried away and go too big to start with.

Begin with something that you can manage and learn with, you can always upgrade it once you are more confident.

The following size system should be more than enough for any beginner to get to grips with.

We can use a simple 1:2 ratio for our flood and drain system.

That is for every 1 gallon (approx. 4 liters) of fishtank water we should use 2 litres of grow bed media.

So a 250 gallon (approx. 1000 liters) fish tank will support 2000 litres of grow bed media.

Fish Tank

The fish tank will most likely be one of the first components of your system that you consider, and as a vital part of your system, you should choose carefully with consideration for water quality, safety, and long-term durability.

It is also likely to be the most expensive components you purchase, so take your time researching the best options. The following points should be considered:

- Shape – most fish tanks are circular, and this is the best shape you can choose. Structurally stronger than other shape, which is useful when you are holding up to and beyond 1000 litres of water, they also have the benefit of better water circulation and flow. This is beneficial for your fish because:

  - A good water flow gives the fish something to swim against, which improves their health and quality of their flesh.

  - Good water flow increases the surface area of the tank, dramatically increasing the oxygen rate.

  - Good water flow prevents layers of protein forming on the surface, which can inhibit the exchange of gases between air and water.

  - In a circular tank, waste will gravitate towards the bottom centre of the tank. Placing your pump in this position will ensure that almost all of the waste is picked up by your pump.

- Materials – most fish tanks are made from food-grade plastic which is the safest option for your system. If you are re-cycling a tank that has another main use, then you should ensure that its material will not compromise the safety of your system in any way and will not leach toxic chemicals into your water. You want your tank to be UV-protected so that the sun does not degrade it over time. As the most expensive part of your system, you want to make sure that it stands up against the elements and lasts for years to come.

- Position – remember that once full a fish tank cannot be moved easily, if at all. Taking time in these early planning stages to purchase the right components, and site them in the best place is the key to a successful aquaponic system.

Grow Bed

There are many possibilities for what to use for the grow bed part of your system.

Chances are that you may want to recycle something that you already have for this component, however, you should take the following into consideration:

- Strength – your grow bed will need to hold a lot of weight when filled with your grow media and water. This can equate to a lot of pressure and so you need to make sure your container is strong enough to cope with this. The grow bed cracking would spell disaster for your system.

- Materials – as with the fish tank, you want your grow bed to be made from food-grade plastic, and be UV protected so as not to leach chemicals into your grow bed and water, or degrade in the sun and weaken the container. You should also ensure your container is waterproof, especially if using a wooden container.

- Position – like the fish tank, once full of grow bed media, it will be very difficult to move. So consider carefully where it is to be sited.

Grow Media

Once you have decided on the size and location of your aquaponic system, you now need to think about the media that you are going to use.

Media is the term given to what you are going to be growing your plants in.

Important points to consider when choosing your media are:

- It must not decompose over time.
- It must not alter the pH level of your water.
- It should be the right size. To small and it will get clogged up with solid waste. To big and the roots of your plants won't be able to get an established hold. Ideally you want your media to be around 12 – 18 mm in diameter.

Good grow medias to use are:

- Expanded clay – pH neutral and light, this is extremely easy to plant in, easy to clean and sterile. The downside is that is can be expensive.
- Expanded shale – pH neutral with rounded edges, makes it easy to use and gentle on plant roots.
- River stone – although heavy, it is easy to handle. However, care should be taken to source it without lime, as this will affect the pH level over time.
- Crushed stone – this is crushed river stone and usually has sharp edges so not user friendly. It also has the same limestone concerns so can affect pH levels over time.
- Synthetic – made from petroleum, it can be very expensive. However, it is light, tends to float, is easy to handle and is pH neutral.

Water Pump

Once of the most important aspects of your aquaponic system will be a water pump to ensure you keep the water moving.

Even just a few short hours of non-moving water will spell disaster for your fish.

The general rule of thumb for getting a pump that is adequate for the job is that you want it to turn over the size of your fish tank in one hour.

So, for our (250 gallon) 1000 liter fish tank, we want a pump that will pump 1000 liters per hour or 1000lph.

Fish

There are many different fish that can be used in your aquaponic system. Factors that will affect your choice are:

- Local climate.
- Availability of stock – bear in mind that you will need to restock periodically.

In order to decide which species of fish will be best for you, consider the following points:

- Are you going to eat your fish?
- If you are going to eat your fish, then how many times a year do you want to harvest them?

The following are all common, hardy choices of fish for aquaponic systems:

Tilapia – an extremely popular fish in aquaponic systems. Easy to breed and fast growing, they can withstand very poor water conditions. They prefer warm water though, and need a water temperature of over 55 degrees F. So if you live in a cold climate they may not be the best choice.

Trout – these are a good choice for cooler climates. They are also fast growing and prefer water temperatures of around 40 degrees F, however, they can tolerate temperatures up to 80 degrees F.

Catfish – there are many different species of catfish around the World and they are well-suited to aquaponic systems. They are very common in the United States, and are quick growing, with a good food conversion ratio. They prefer temperatures of around 70 degrees F.

Yellow Perch –is a very adaptable fish and are another species that are fast growing. They prefer water temperatures from 65 – 70 degrees.

Goldfish – are the perfect choice if you are not going to be eating your fish. They are pretty tough and easy to breed, although they generally need plant cover in order to do so. They prefer temperatures between 68 – 75 degrees F.

Other species for your aquaponic tanks are:

- Fresh water mussels.
- Fresh water prawns and crayfish.

Mussels are a filter feeder and will help clean the water. They will happily grow in either the flooded grow bed or in the fish tank.

Crustaceans are a nice addition but this choice will depend on water temperature and your location and availability.

Stocking Levels

Fish stock – An average rule of thumb is 1lb of grown fish weight will require 5 – 7 gallons of water. Be sure you know the end growth weight of your fish when you purchase them.

In our case, a 250 gallon (1000 liter) fish tank will support around 40 fish (depending on size) or around 40lb of fish.

Culling Your Fish

Your fish will grow at different rates, which is a good thing as you will not want to harvest them all at the same time.

You want to achieve an ongoing cycle of young fish that are still growing, and older fish that are ready to eat.

Be mindful of your stocking levels and make sure that you don't allow your tank to become overcrowded. This leads to an in balance in your system, which in turn can affect the health of your fish.

How to Cull You Fish

When you are ready to cull your fish, you should first identify which ones you want to cull. Using a net, scoop them out of the tank in an easy and smooth motion. There is no need to stress your fish out by thrashing the net around in the water.

Once out of the water, some fish, such as the catfish, can live for several hours. During this time the fish will be producing adrenaline and other chemicals which move through the flesh. This is not desirable as you do not want to be eating this, and it is not a pleasant death for your fish.

It is best to use to bleed the fish that is cut its throat with a sharp knife.

Wear thick leather gloves to do this, which help ensure you can keep a firm grip on the fish, and prevent any accidents from the knife.

Hold the fish firmly in one hand, remember your fish will be flipping around, and make a swift and firm cut to its throat. Wash and clean the fish as normal.

What Do I Feed My Aquaponic Fish

Fish are omnivorous and require a range of protein in their diet, however, some will need more protein than others. You should be aware of the diet of your fish, and if you have a mixed range of fish species in your tank, you should make sure that you cover the diets of all your fish adequately.

The basic food for your fish should be a pellet food, which will provide them with all the nutrition/and protein that they need.

You can supplement with alternate food sources such as:

- Worms.
- Maggots.
- Larvae.
- Plant bugs and insects (from your plants).

In a mature system you can feed your fish once or twice a day. Feed them as much as they will eat for a few minutes at a time (preferably later in the day), and remove any uneaten food from the surface after 30 minutes, otherwise it will sink and rot.

Organic Fish Feed

Try wherever possible to feed your fish organic fish pellets. Remember the adage "you are what you eat", whatever you feed your fish, ultimately you will be eating.

In which case you want to make sure that your fish feed is non-GMO, non-soya, has no land based animal components in it and is also free from fish meal.

Buying from a reputable source is the best place to start when first starting out with fish.

Organic fish pellets will ensure your fish are getting adequate nutrition, although in time you may wish to move on to making your own fish pellets.

Starting Your System

Cycling

Once you have built your system, filled your grow beds with the media of your choice, and filled your fish tanks with de-chlorinated water, you are ready to populate your system.

In order for aquaponics to work its magic, you need to establish a beneficial colony of bacteria.

This process is often called cycling and begins by adding ammonia to the system. This can be done one of two ways:

- Cycling with fish.
- Cycling without fish.

Ammonia is extremely toxic to fish and will kill them unless it is diluted or converted into a less toxic form of nitrogen.

The process of conversion works due to the presence of two forms of bacteria.

Firstly, the ammonia attracts the bacteria 'nitrosomonas' which converts the ammonia into nitrites. These nitrites are even more toxic than the ammonia, but attract the second bacteria 'nitrospira'.

This is the magic ingredient that will convert the nitrites into nitrates, which are generally harmless to fish and great food for your plants.

It is important that you continually test the water during this time to see where you are in the process.

You will need to test your ammonia, nitrite and nitrate levels as well as your pH levels.

Once you can detect nitrate in your water and both ammonia and nitrite levels have fallen to below 0.5ppm your system will be fully cycled.

Cycling With Fish

Setting off the cycling process using fish will typically take around 4 – 6 weeks.

Adding your first fish will set the process off. It is recommended that you don't fully stock your tank at this point, as it is possible that you will lose some fish during the process.

Stock to half capacity and use goldfish if you can as they are more tolerant to ammonia.

Only feed your fish once a day during the cycling process and with a small amount of feed only.

Ammonia levels

Fish prefer slightly alkaline water and you should try to keep the pH level between 6 – 7 whilst cycling.

Below is a chart of pH levels and levels of ammonia measured in mg/l (ppm) that can be considered safe at a given temperature and pH level.

pH	20c (68f)	25c (77f)
6.5	15.4ppm	11.1ppm
7.0	5	3.6
7.5	1.6	1.2
8.0	0.5	0.4
8.5	0.2	0.1

Monitor your levels daily. If the levels exceed those on the chart then you will need to dilute the water by pumping out ⅓rd of the water and replacing it with fresh de-chlorinated water.

pH Levels

During cycling you will need to keep your pH levels within a tight range of 6 – 7

Although it is vitally important to keep to this range, it is equally as important to make any changes slowly. Fast and large pH swings can be extremely stressful to your fish, even more so than having a pH out of range.

The advice is to make any changes at 0.2 increments per day.

The safest way to do this is by using special kits that will raise and lower your pH as required.

Alternatively you can use calcium hydroxide, also known as "builders lime" or "hydrated lime", together with potassium carbonate/bicarbonate or potassium hydroxide.

However, for beginners, the pre-made kits may be the better option.

During cycling you will find that you are mainly trying to keep the pH levels down. Once cycling is complete you will find that you are switching to keeping them up.

The ideal pH of an aquaponics system is between 6.8 – 7.0

Nitrite Levels

Nitrite to fish is like carbon monoxide to air breathers. It binds with the blood and keeps the fish from getting oxygen.

If at any time during the cycling process the nitrite level goes above 10ppm then you should do a water exchange as discussed above.

You can also add non-iodized salt to your water to help improve the levels. Dilute at 1 part per thousand (1kg per 1000 liters of water). Do not use table salt as this contains caking agents. You can use pool salt, water softener salt or specially formulated salt for fish.

Dilute completely in a bucket of water before adding to your fish tank. Any crystals that sink to the bottom could potentially burn your fish if they rest on them.

Stop feeding your fish until the nitrite levels have dropped below 1.0 ppm.

Testing kits to test each of the levels discussed are readily available from numerous sources: gardening outlets, aquatic centers or online.

The Importance Of Testing Tools

Testing your water throughout the cycling process is the only way of knowing if your system is cycled and ready to go.

Pre-made up kits to test each of the levels discussed are readily available from numerous sources: gardening outlets, aquatic centers or online.

You will be testing:

Ammonia levels/nitrite levels/nitrate levels - one of three ways.

- Strip test – simply dip the correct end in the tank water.
- Powder test – pour tank water into a test tube and add the powder.
- Liquid test – same as above but using a liquid and not a powder.

With all the above tests you will be comparing the color of the results with a chart given in your kit, to see what the ranges are.

pH levels

Test pH levels with a simply litmus paper test. Dip the end into the tank water and compare the color with the color chart provided in your test kit.

Cycling Without Fish

The alternative to starting your cycling process with fish is the fishless method.

It has some major benefits over cycling with fish:

- Much quicker process – only takes between 10 days and 3 weeks, due to elevating the ammonia too much higher levels than you would with fish.

- Less stress involved – you need to be less concerned with pH levels as you are when you have to keep your fish alive.

- Better bacteria – you will typically find that the bacteria is more robust using this method.

- Fully stocked tank – you can fully stock your tank with fish when you are finished, unlike the more gradual approach recommended when you cycle with fish. This is especially beneficial if stocking with carnivorous fish, as they are less likely to eat each other if introduced together.

- Control – you can control exactly how much ammonia is added to your system.

There are several options for adding ammonia to your system. These are:

- Liquid ammonia – only use pure 100% ammonia. Avoid anything with perfumes, colorants or additives.

- Crystallized ammonia – as above but in a crystal form.

- Human urine – an excellent form of ammonia and no cost involved! Needs to be left in a sealed bottle for a week or two to allow the urea to convert to ammonia.

How to Fishless Cycle

- Add your source of ammonia a little at a time until you get a reading of 5ppm.

- Make a note of how much ammonia this took, and then add this amount daily until your nitrite level reads 0.5ppm.

- Once nitrites appear at this reading, cut back the ammonia by one half and continue until you get a nitrate reading.

- When your nitrite levels have dropped to zero, you can add your fish.

Growing Your Produce

Once you have successfully added your fish, you can start to grow your produce.

Root vegetables do not do so well with this form of gardening, although some people have had success with them.

Most other vegetables do very well within an aquaponic system and you should have great success with:

- Green leafy vegetables: all types of lettuce, kale, spinach, bok choi, pak choi.
- Tomatoes.
- Beans and peas.
- Peppers and chillies.
- Eggplants.
- Zucchini (courgette).
- Cruciferous vegetables: cabbage, broccoli, cauliflower.
- Pumpkin and other squashes.
- Melons.
- Strawberries and blueberries.
- Rhubarb.
- Some fruit trees.
- Herbs – all varieties.

Seeds or Seedlings

Both seeds and seedlings can be used within your system and you should aim to grow a mix of the two.

In fact, it is a good idea to plant some seedlings into your grow bed at the beginning of your cycling process. It can take a little while for the roots to become established and start to use the fish effluent, so you should aim to begin this process as soon as possible.

The use of a good organic seaweed based growth stimulant can be used here to speed up the process.

Germinating Seeds

The general rule of thumb here is that smaller seeds can be scattered evenly over your grow bed. These seeds will fall through your media and reach the appropriate depth in which to germinate.

Small seeds would include: lettuce, green leafy vegetables, tomatoes, and seeds that are usually planted out in early spring, as these can cope with a wet environment in which to germinate.

Larger seeds such as: beans, cucumber, eggplants, melons and "longer term" plants are best germinated before adding to your grow bed.

Start them off on a piece of wet paper towel and either place in an air-tight container or a Ziplock bag.

Check on them daily and when you have a root of 25mm you can transplant them to your grow bed, making sure you place them at a level where they will become wet with the water cycles.

Seedlings

Alternatively, you may wish to simply use seedlings that you have brought.

Simply shake off the excess dirt and run the roots under a gentle tap to wash. Free them from any bugs there may be and plant as above.

Spacing

Plants can be grown much closer together in an aquaponic system because they are not in competition with each other for nutrients. There is more than enough for everyone.

Whilst you are able to grow plants twice as close as you would in traditional planting, you want to make sure you leave enough room for air circulation, light and to control insects and bugs.

A good rule of thumb is to think about the finished size of your plant – i.e. a head of lettuce, and then give just enough room for it to grow to that size before touching on the space of the next plant.

Insects and Bugs

Pesticides should not be used in an aquaponic system as they will end up in the water that is being sent back to your fish tank and ultimately could kill your fish.

Due care should be taken with any organic solutions that you use so as to avoid them getting in the water supply.

It is best therefore to avoid using any sort of sprays at all to control insects and bugs on your plant.

Natural methods of insect control work just as well as long as you are vigilant. Try the following methods:

- Spray any insects/bugs that you see with water. This will just wash them off the plant and into the water that is recycled into the fish tank. Once in the fish tank the fish will eat the insects/ bugs.

- Slugs, larvae or caterpillars can be collected and fed to your fish, who will appreciate the extra "snacks".

- Leave saucers of beer to attract slugs or leave crushed egg shells around your grow bed to prevent slugs and caterpillars from crossing into your grow bed.

- You can also introduce some insect "predators" to control any pests that you may have. Ladybirds are a particularly useful insect to use as a "pest controller" and will soon eat any aphids that may be attacking your plants. You can also buy praying mantis eggs that once hatched also make a very effective pest controller.

Troubleshooting

Given that you have gone to so much trouble to set up your aquaponic system, it would be a real shame if you now failed because of a problem that you can resolve very easily.

Below are 5 of the most common problems that people come across when dealing with aquaponics:

1. Bug problems – we have already dealt with this one, but the most important thing is to deal with them as soon as they appear so the problem doesn't get out of control. Feed slugs, larvae and caterpillars to your fish, spray larger infestations with water, and use ladybirds as an effective "pest controller".

2. Water temperature – controlling the water temperature in your tank can save you a lot of problems. It is important that you know the ideal water temperatures of your fish, and be especially vigilant during hot, sunny weather. Hot water means less oxygen available for your fish, which potentially equals dead fish.

3. Overcrowding – do not allow the population of your fish to become too dense as this can result in:

   • Larger fish feeding on the smaller fish.
   • Too much effluent for your system to handle.

4. Too much ammonia – you should test your water regularly (once a week), to ensure the water is not becoming too toxic to your fish.

5. Algae in the fish tank – algae will feed your water with oxygen during the day and starve it of oxygen at night (not good for your fish). It will also raise the pH level of your water in the day and lower it at night, causing fluctuations that your fish will not like. The simplest way of protecting your tank from algae is to keep it covered from the sun. This will also help you maintain an even water temperature.

Conclusion

Aquaponics is a super-efficient way of growing both produce and fish.

Even if you don't intend to use it as a way to raise and cull your own fish, the benefits you will gain from growing produce alone, are worth it in their own right. You can produce over 4 times as much produce as traditional gardening would give you.

A flood and drain system like the one we have shown in this book, is an easy and affordable way to get started with aquaponics, and in no time at all you will be harvesting tasty and plentiful produce.

With no waste and minimal maintenance work required, aquaponics is fast becoming a very real and sustainable option for home growing.

It uses as little as 2% of the water that conventional gardening takes, saves on fertilizers, pesticides and herbicides, and cuts out the need for fertile soil.

Yields that are higher and growing times that are shorter make this a very attractive and affordable step towards becoming self-sufficient.

Happy growing!

Brian Grant

Thank you for reading "*Aquaponics Made Easy: A Simple and Easy Guide to Raising Fish and Growing Food Organically in Your Home or Backyard*"!

Sign up for updates of our new books, free bonuses and more...

www.SparrowPublications.com

Review

If you have any questions or comments, feel free to email me at help@sparrowpublications.com. I do my best reply to all questions that come in and that I am able to.

Enjoy this book? Please share a couple of sentences and a 4-5 star review on Amazon. It would mean a great deal to me and others who are looking.

Credits

Diagram from page 5 sourced from:
http://theaquaponicsource.com/what-is-aquaponics/

Diagram from page 7 sourced from:
http://homeaquaponicssystem.com/diy/aquaponics-system-design-flood-and-drain/